Kareem Khalil

Liquid Marbles

Kareem Khalil

Liquid Marbles

Granulation of Liquid Marbles of Different Fluid Properties Recorded by High Speed Imaging

LAP LAMBERT Academic Publishing

Impressum / Imprint

Bibliografische Information der Deutschen Nationalbibliothek: Die Deutsche Nationalbibliothek verzeichnet diese Publikation in der Deutschen Nationalbibliografie; detaillierte bibliografische Daten sind im Internet über http://dnb.d-nb.de abrufbar.

Alle in diesem Buch genannten Marken und Produktnamen unterliegen warenzeichen-, marken- oder patentrechtlichem Schutz bzw. sind Warenzeichen oder eingetragene Warenzeichen der jeweiligen Inhaber. Die Wiedergabe von Marken, Produktnamen, Gebrauchsnamen, Handelsnamen, Warenbezeichnungen u.s.w. in diesem Werk berechtigt auch ohne besondere Kennzeichnung nicht zu der Annahme, dass solche Namen im Sinne der Warenzeichen- und Markenschutzgesetzgebung als frei zu betrachten wären und daher von jedermann benutzt werden dürften.

Bibliographic information published by the Deutsche Nationalbibliothek: The Deutsche Nationalbibliothek lists this publication in the Deutsche Nationalbibliografie; detailed bibliographic data are available in the Internet at http://dnb.d-nb.de.

Any brand names and product names mentioned in this book are subject to trademark, brand or patent protection and are trademarks or registered trademarks of their respective holders. The use of brand names, product names, common names, trade names, product descriptions etc. even without a particular marking in this works is in no way to be construed to mean that such names may be regarded as unrestricted in respect of trademark and brand protection legislation and could thus be used by anyone.

Coverbild / Cover image: www.ingimage.com

Verlag / Publisher:
LAP LAMBERT Academic Publishing
ist ein Imprint der / is a trademark of
OmniScriptum GmbH & Co. KG
Heinrich-Böcking-Str. 6-8, 66121 Saarbrücken, Deutschland / Germany
Email: info@lap-publishing.com

Herstellung: siehe letzte Seite /
Printed at: see last page
ISBN: 978-3-659-60955-8

ABSTRACT

Liquid Marbles

Kareem Khalil

Granulation, the process of formation of granules from a combination of base

powders and binder liquids, has been a subject of research for almost 50 years,

studied extensively for its vast applications, primarily to the pharmaceutical

industry sector. The principal aim of granulation is to form granules comprised

of the active pharmaceutical ingredients (API's), which have more desirable

handling and flowability properties than raw powders. It is also essential to

ensure an even distribution of active ingredients within a tablet with the goal of

achieving time-controlled release of drugs.

Due to the product-specific nature of the industry, however, data is largely

empirical [1]. For example, the raw powders used can vary in size by two orders

of magnitude with narrow or broad size distributions. The physical properties of

the binder liquids can also vary significantly depending on the powder

properties and required granule size.

Some significant progress has been made to better our understanding of the

overall granulation process [1] and it is widely accepted that the initial

nucleation / wetting stage, when the binder liquid first wets the powders, is key

to the whole process. As such, many experimental studies have been conducted

in attempt to elucidate the physics of this first stage [1], with two main

mechanisms being observed – classified by Ivenson [1] as the "Traditional

description" and the "Modern Approach". See Figure 1 for a graphical definition

of these two mechanisms.

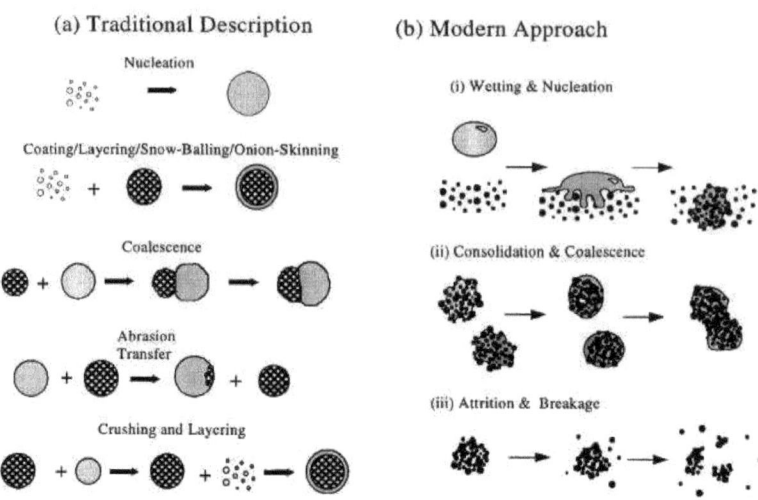

Figure 1: Traditional vs. Modern Approach [1]

Recent studies have focused on the latter approach [1] and a new, exciting development in this field is the *Liquid Marble*. This interesting formation occurs when a liquid droplet interacts with a hydrophobic (or superhydrophobic) powder. The droplet can become encased in the powder, which essentially provides a protective "shell" or "jacket" for the liquid inside [2]. The liquid inside is then isolated from contact with other solids or liquids and has some fascinating physical properties, which will be described later on. The main potential use for these liquid marbles appears to be for the formation of novel, hollow granules [3], which may have desirable properties in specific pharmaceutical applications (e.g. respiratory devices). They have also been shown to be a highly effectively means of water recovery and potentially as micro-transporters and micro-reactors [4].

However, many studies in the literature are essentially proof-of-concept approaches for applications and a systematic study of the dynamics of the marble formation and the first interactions of the liquid droplet with the powder is lacking. This is the motivation for this research project, where we aim to provide such information from an experimental study of drop impact onto hydrophobic powders with the use of high-speed imaging.

ACKNOWLEDGEMENTS

This work was funded by Professor Sigurdur Thoroddsen and was performed in his laboratory (High Speed Fluids) at King Abdullah University of Science and Technology. I highly appreciate the support and contribution of Dr. Jeremy Marston for his efforts in the work overall; experiments, results, analysis and outputs. In addition, I would like to thank Dr. Erqiang Li who helped develop the experimental setup and protocols. Finally, I would like to thank Dr. Ivan Vakarelski for his help in preparing the hydrophobic powder.

TABLE OF CONTENTS

Organization of Thesis

This thesis is organized as follows: In the following section a nomenclature is provided which is followed throughout the thesis. Chapter One provides an in-depth introduction into the topic of granulation and previous studies on the first stages of the granulation process. A review of studies specifically related to hydrophobic powders and liquid marbles is also given in Chapter Two. The aims and objectives are summarized in Chapter Three. Chapter Four represents the experimental setup used, with details of the camera, powder characterization and liquid physical properties. In Chapter Five, some qualitative examples of the dynamical process are explained, showing the main qualitative features of the impact-rebound process and marble formation (In addition, some example video clips are provided on the CD/USB drive attached with this thesis). In the same chapter, the quantitative data and main results and discussion are provided. Finally, Chapter Six draws conclusions and discusses directions for future work in this area.

LIST OF SYMBOLS

g: Acceleration due to gravity.

h: Distance of the droplet penetration under the powder surface.

r: Radius of the droplet.

k: Thickness of the droplet standing on top of the powder.

l: The contact length of the droplet with the powder surface.

A: The area of the base of the droplet, which is in contact with the surface.

D: Original diameter of the droplet before impact.

D_{max}: Maximum Diameter of a drop during the spreading phase, measured at the equator.

u_i: Impact velocity.

d: Instantaneous drop diameter during spreading / rebound.

d_b: Powder particle diameter.

We: Weber number.

Bo: Bond number.

Oh: Ohnesorge number.

  : Viscosity of the fluid.

  : Density of the fluid.

κ: Capillary length.

  : Surface tension of the fluid.

  : Ratio of the solid/liquid interface area and the liquid surface area.

  : Contact angle between the droplet and the powder.

  : Packing fraction of the powder.

t: Time instants during the experiments.

t_i: Impact time, which is the time from touching the powder till leaving it.

t_r: Rebound time, which is the time in the air from leaving the powder till touching it again.

V_0: Volume of the droplet.

r_d: Radius of the droplet footprint on the powder surface.

z: Rebound height.

x: Rebound length; which is the distance in the x-direction for the rebounding droplets from inclined surfaces.

α: Degree of inclination of the target powder surface.

: Surface porosity.

R_{pore}: Effective pore radius based on cylindrical pores.

p: pressure inside the droplet.

P: The difference between the pressure inside the droplet and the atmospheric pressure.

: Spreading coefficient.

$_{L/S}$: Liquid spreading over solid coefficient.

$_{S/L}$: Solid spreading over liquid coefficient.

$_{SV}$: Surface free energy between the solid and the vapor states.

$_{LV}$: Surface free energy between the liquid and the vapor states.

$_{SL}$: Surface free energy between the solid and the liquid states.

$_{S}^{d}$: Surface free energy for the solid with non-polar intermolecular interactions.

$_{L}^{d}$: Surface free energy for the liquid with non-polar intermolecular interactions.

$_{S}^{p}$: Surface free energy for the solid with polar intermolecular interactions.

γ_L^P: Surface free energy for the liquid with polar intermolecular interactions.

W_{CS}: Work of cohesion for solid.

W_{CL}: Work of cohesion for liquid.

W_A: Work of adhesion.

LIST OF FIGURES

LIST OF TABLES

Chapter One: Literature Review

1.1   Overview of the Granulation Processes

Granulation has many applications in the pharmaceutical industry, minerals and fertilizer granulation. The top four companies that produce pharmaceuticals worldwide according to the order of ranking in terms of market capitalization (total number of shares in issue * share price) are Johnson & Johnson, Pfizer, Roche and GlaxoSmithKline. According to Wikipedia [5], the total revenues for Johnson and Johnson, for instance, is around 62 billion dollars annually. The net income for Pfizer is about 8.6 billion dollars. Annually, GlaxoSmithKline spends 4.1 billion British pounds for research and development. Given that pharmaceuticals are the main product of these companies, one can understand the need for studying processes such as granulation in detail.

In addition, other fields that benefit from the study of drop impact onto granular layers are planetary science, material science, civil engineering and agriculture [6]. 20% of manufactured products in the chemical industry use powder as ingredients [1], thus it is important to study wet granulation – i.e. the result of mixing fluids with powders.

Specific to the pharmaceutical industry, the mixing of powders with fluids can help produce a form of an active pharmaceutical ingredient (API) that is easier to handle and implement in the tableting process. Changing certain parameters, such as the powder size or liquid physical properties, can lead to different distributions of API's in the tablets. Katsuragi [6] illustrates that granulation as a process can be helpful in understanding the scientific methodology and the novel behind the impact of projectiles from the geological point of view and can be helpful in understanding the phenomena associated with rain droplets falling on different surfaces.

The impact of droplets onto powder surfaces is the first step in the "modern approach" to granulation [1]. Largely, improvements to granulation are as a result of an empirical process due to the specific nature of the end products. The first stage of granulation is the wetting or nucleation stage when the fluid starts to touch the powder. The region around an impacting binder droplet can be considered the "Nucleation" or "Wetting" zone. Nuclei are then formed inside the powder. The binder dispersion, which is the second step, is a function of process variables [1], such as $, D, , , .$ For instance, the impacting droplet size can span three orders of magnitude, which in itself can result in multiple outcomes for the impact. Other parameters are the change in impact velocity and the packing fraction. In addition, one must consider the physical properties of the liquid used – namely – the viscosity, surface tension and density. Using these properties and process variables, it is possible to form several dimensionless numbers that can help quantify the regimes of post-impact phenomena for decision making. For instance, in the case of splashing, it is found that the Weber number is the governing dimensionless parameter for drops with Bo>1. Ivenson [1] states that the two most important factors in the binder dispersion within the powder by order are the contact angle and the solid to liquid ratio. As the contact angle increases (e.g. in the case of having a more hydrophobic powder), the wettability of the powder decreases. The powder wettability is generally quantified by means of an apparent contact angle of the liquid with the powder surface, i.e. a drop which has high apparent contact angle means that the powder has a low degree of wettability. The argument presented by Ivenson [1] that the contact angle is the most pertaining to successful granulation is sound

because the contact angle can be considered the umbrella for several other

parameters, since it is a direct function of several parameters, i.e.

$$\theta = f(\mu, \quad, \sigma)$$

As such, changing one of these basic properties would subsequently change the

contact angle. The second most important parameter according to Ivenson [1] is

the spreading coefficient , which is a measure of the tendency of a liquid to

spread over a solid or vice versa and is related to the work of adhesion and

cohesion, as follows:

$$_{L/S} = W_A \quad W_{CL}$$
$$_{S/L} = W_A \quad W_{CS}$$

According to Ivenson [1], the difference between the work of cohesion and the

work of adhesion is the spreading coefficient .

Work of cohesion for a solid: $W_{CS} = 2 \quad_{SV}$

Work of cohesion for a liquid: $W_{CL} = 2 \quad_{LV}$

Work of adhesion for an interface:
$$W_A = \quad_{SV} + \quad_{LV} \quad_{SL}$$
$$W_A = \quad_{LV}(\cos \quad +1)$$

Ivenson states that $_{L/S}$ is positive when the droplet spreads over the powder

and form a layer and it is more positive if it forms a strong granule inside the

powder. Conversely, if $_{S/L}$ is positive, a liquid marble is formed. There is a third

case where both liquid and solid have high works of cohesion, so the solid liquid

interfacial area is minimized and this case represents non wetting [4].

The time and length scales of penetration are also an important consideration for the granulation process, as recently studied by Katasuragi [7], showing that the timescale of impact onto a powder is dependent on the liquid viscosity, but is independent of the impact speed, which will be discussed later.

The powder surface itself acts sometimes as a fluid and sometimes as a solid. According to Katasuragi [7], it behaves as a fluid at the instant of the impact of the droplet on the powder surface, and it behaves as a solid later during the stable state of the droplet. Many other studies exist in the literature concerning the fluid-like response of powders to impacts [7,8].

1.2 Nucleation and Binder Spray Addition

In general, there are two cases for the nucleation; either the droplet size is comparable to the base powder (particle) size, or the droplet size is too large compared to the powder size. These two cases are shown graphically in Figure 2. In the first case, when the droplet size is small, then the droplets coat the particles, which then coalesce on the surface of the powder bed [1]. In the second case, the particles immerse in the liquid droplet, which can also result in the generation of saturated pores.

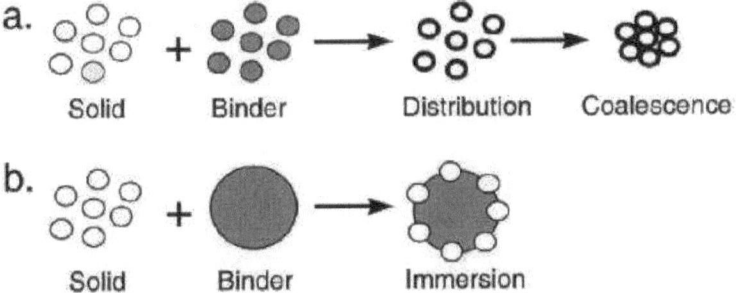

Figure 2: Two cases of binder dispersion where (a) drops sizes are close in size to the powder particles and (b) where the drop size is larger than the base powder size. The figure is reproduced from [19}.

1.2.1 Definition of Powder Wettability

The final distribution of API's in granules and thus the tablet largely depends on the granulation process where the contact angle plays an important role [4]. The contact angle is the angle formed between the droplet of the fluid and the powder in contact with the fluid. If this angle is greater than 90, then the powder is considered to be hydrophobic and, generally liquid marbles are formed. If the angle is between 70° and 90°, then the wetting according to Hapgood [4] is poor, where it was noted that in order for the granulation to be successful, wetting must occur. In case of using a hydrophilic powder the droplet penetrates the powder.

1.2.2 Timescales

There are two important scales that characterise the impact process; the contact time, t_i, also referred to as the impact or deformation time, and the rebound time, t_r, also referred to as the oscillation time. The impact time (t_i) is the total time from when the drop first contacts the powder surface until it leaves the surface and starts rebounding. The rebound time (t_r) is the time when the droplet leaves the surface after impact or deformation till it touches the powder again. In other words, it is the duration of the droplet in the air. The rebound / oscillation time was found by Katsuragi [7] to be in the range 0.01 – 1 seconds for different fluids such as water, glycerol and ethanol. He used in his experiments SiC and glass beads with diameters of 4 µm and 50 µm respectively. He also reported in another publication [6] that the deformation time is constant for a given fluid and is of the order of 10^{-2} seconds for most fluids tested therein. It thus appears to be independent of the properties of the powder / bed preparation. A similar observation was made by Clanet et al. [9].

1.2.3 Other Considerations for Binder Sprays

There is an established relation between the droplet diameter and the granule diameter, generally taking the form

$$d_b \, \alpha \, D^n$$

Where d_b is = powder/base particle diameter and d_g is used for granule. Where n is a correlation coefficient. This correlation coefficient depends on the fluid and powder properties. As an example, water and lactose powder with d_b = 100 m have a correlation coefficient of n=0.89, which appears to hold for a range of $35 \le D \le 3$ m [1].

The liquid flow rate to the binder spray also clearly affects both the binder drop sizes and the nucleation rate [1].

Another parameter for the degrees of freedom after the granule diameters and the flow rate is the nozzle/needle position from the impact surface. Large spray angles and increased nozzle heights above the beds decrease the likelihood of grouping or coalescing of droplets and thus increase the rate of successful nucleation [1].

Figure 3: Wet granulator [11]

Figure 3 shows a wet fluid-bed granulator. It is used for spraying droplets over the powder, and so the results are binder droplets.

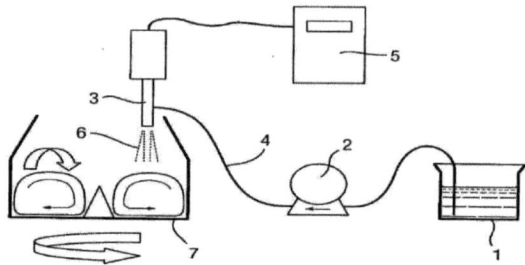

Figure 4: Schematic of wet granulator, where 1 – 7 denote the fluid, a pump, a nozzle, a pipe, the feeding device, fluid droplets and a circular mixer respectively [12]

The fluid is fed from a holding tank as shown in Figure 4. The pump pushes the fluid through a pipe to the nozzle where the feeding device controls the rate of feeding the micro droplets. The droplets fall over the solid surface to form nucleation sites and, eventually, granules. A circular mixer at the bottom rotates in order to distribute the liquid binder. The next step is the moistening in which the liquid bridges are formed between the binder droplets. Then the bridges solidify. Finally, the finished agglomerate can form a shape like a snowball structure. These steps are shown in details in Figure 5.

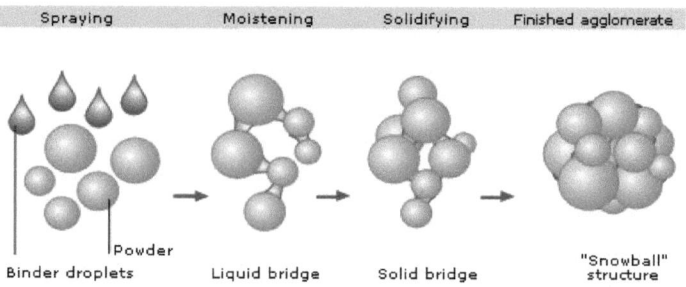

Figure 5: Granulation steps

Chapter Two: Hydrophobic Powders

A hydrophobic powder is defined as a powder upon which a liquid drop forms a contact angle of greater than 90°. Generally, they are produced by chemically pre-treating the powders (discussed below). This will inevitably lead to poor wetting from the nucleation perspective. In the case of droplets impacting onto a hydrophobic powder, the droplets deform, reach a maximum spread and then either roll or rebound from the surface of the powder bed. Hydrophobic components in drug manufacturing are of extreme importance. Traditional reasons for the inclusion of hydrophobic powder were to improve flow and handling properties, increase bulk density, and reduce segregation of materials. Classical methods of granulation were introduced to improve the quality of drugs in general. In addition, granulation nowadays helps in fast, mass-transfer limited drying properties, good particle rearrangement and compression characteristics for tableting in the case of forming hollow granules. Besides, it is easy to obtain a reproducible dissolution. Most importantly, there is a potential to load a solubale drug in the fluid interior and granulate with a second hydrophobic drug to form the outer shell [4].

The contact angles between a hydrophobic powder and a given fluid have constant values in the case of a static state of droplets without motion. For instance, the following table identifies some values for the contact angles between water and different kinds of powders.

Powder type	Contact angle (deg)
PTFE	140
OTFE	177
Aerosil R972	103
Aerosil R812S	113
Aerosil R202	>113
Hydrophobic copper powder	157
Poly-methylmethacralate (PMMA)	120
Lycopodium powder	>150

Table 1: Contact angle values for several powders with water [13]

2.1 Treatment to Produce Hydrophobic Powders

Recent studies have shown several ways of treating powders in order to produce hydrophobic or super hydrophobic powders. According to the research conducted by Quere et al. and Hapgood et al., the following table summarizes the powders used in their research and how they are treated to be hydrophobic in

Reference	Powder material	Grain size (μm)	Grain density (g/cm^3)	Treatment (chemical agent)	Resulting water contact angle (deg)
Quere et al.	Lycopodium	30	unknown	Flouro-decyl-trichlorosilane	160-165
Quere et al.	Silica	0.01	1.5	Dichloro-demethyl-silane	113
Hapgood et al.	Glass ballotini	65, 121 and 191	2.5	Sigmacote	unknown
Hapgood et al.	Polytetra fluoroethylene	1, 12, 35 and 100	2.1	No treatment; ready made	140

some cases.

Table 2: Reported methods for treatment of powders [11, 12]

2.2 Impact Dynamics

The behavior of liquid droplets impacting on hydrophobic powder is dependent on the fluid properties and the pre-treatment of the powder itself. Several non-dimensional numbers can incorporate the influence of these individual properties, such as the Bond number, Weber number and Ohnesorge number. These numbers help to identify thresholds between some of the observed behaviors such as rebounding, spreading and splashing. These three dimensionless numbers for the fluid are defined as

$$Bo = \frac{_L g R^2}{} \qquad We = \frac{U^2 D}{} \qquad Oh = \frac{}{\sqrt{} D}$$

These three numbers give the ratios of gravity to surface tension, inertia to surface tension and viscosity to surface tension, respectively. When a liquid drop impacts onto a solid (or powder surface), it can deform into a pancake-shaped, more commonly referred to as a 'puddle' [9]. This observation is shown graphically in Figures 6 and 7.

Figure 6: Droplet just before impact [11].   Figure 7: Droplet at the instant of impact forming puddle [11].

The volume of the droplet before the impact is given by: $V = \dfrac{4}{3} r^3$. The volume of the droplet in Figure 7 is given by: $V = h \times \dfrac{D_{max}^2}{4}$. The puddle height is given by: $h\alpha\sqrt{\dfrac{\sigma}{\rho g}}$. It was also noted that the characteristic timescale for the impact is $t_i = \dfrac{D}{U}$, so that the typical acceleration felt by the drop is $\gamma = \dfrac{U^2}{D}$. Using this modified acceleration in lieu of gravitational acceleration, we then have $h = \sqrt{\dfrac{\sigma D}{\rho U^2}}$. Also, by volume conservation, $D_{max}^2 \quad \dfrac{D^3}{h}$, so that we have the relation $D_{max}^2 \quad \dfrac{D^3}{\sqrt{\dfrac{D}{U^2}}}$, which then yields the following expression for the maximum diameter in terms of the Weber number $D_{max} \quad D.We^{1/4}$.

A plot of the spread diameter, d, verses time, t, where both axes have been normalized using the length scale, D, and time scale, D/U, respectively. Once the droplet touches the surface, the diameter increases until it reaches the maximum

of the spread indicated by the maxima on each curve. As the drop recoils, the

ratio approaches 1 slightly as it approaches a final state.

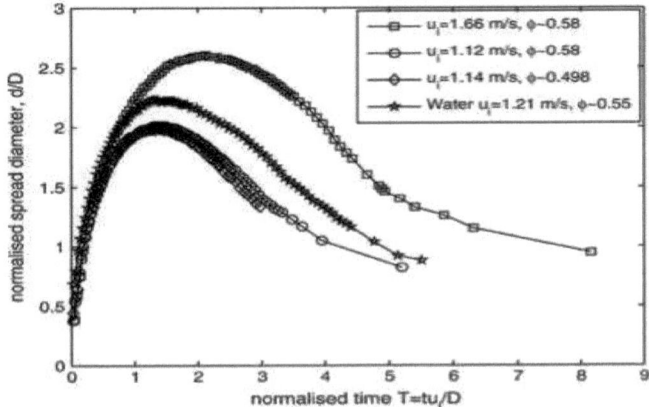

Figure 8: Relation between the time and the maximum spread at different parameters [11]

Four cases are shown on the graph. Three of them are for 50% glycerol. One

case is for pure water. The highest value for the maximum spread happens at the

highest impact velocity. At three cases out of the four, $d/D \approx 1$, which means that

the contact diameter of the drop after motion has ceased is equal to its initial

drop diameter before impact except the one at the lowest packing fraction, in

which this shows that the more loosely the bed is packed, the less possibility the

droplet returns to its original shape. Figure 9 (a) shows the maximum spread of

acetone, ethanol and 50% ethanol drops, plotted against the impact velocity,

showing a linear increase in D_{max} until $U\sim2$ m/s, at which point the drops shatter

upon impact (splashing) producing satellite drops. The corresponding Weber

numbers for this threshold are 20. Snapshots showing the maximal deformation

at three points in the data set are shown below the graph. Part (a) corresponds to the first snapshot. Part (b) corresponds to the second snapshot and Part (c) corresponds to the third snapshot, in which the spread of the drop cannot be predicted.

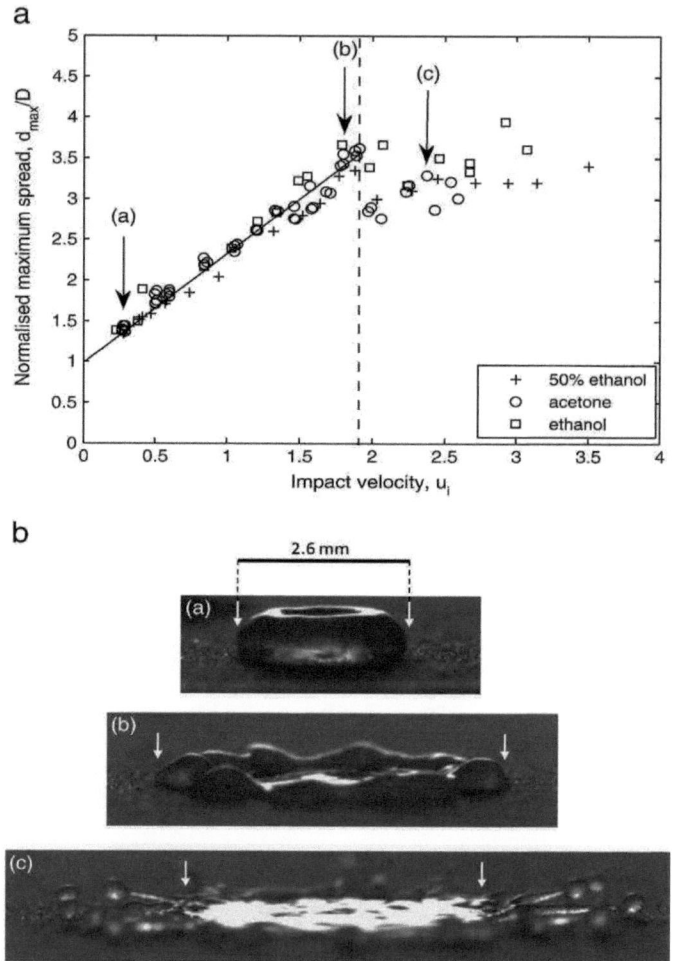

Figure 9: (a) Maximum spread diameter (normalized by initial drop diameter) plotted against impact velocity ui, for liquids with B0>1. The solids packing in the base varied slightly between repeat trials in the range 0.51<φ<0.56. The dashed line marks the onset of satellite drop detachment from the fingers. (b) Images of acetone drops at maximum spread in different regimes, indicated by arrows on the plot in part (a) [11].

After the droplets impact on the hydrophobic powder, they form quasi-spherical shapes such as that shown in Figure 10 , which are largely spherical except that at the bottom, their bases are flattened under their own weight [14].

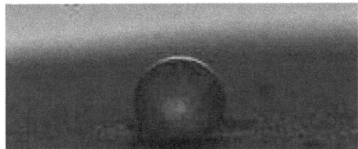

Figure 10: Liquid marble at rest [10]

There are two different shapes for liquid marbles, after the droplets have been coated with powders [14]. Either they deform, squeeze and form a puddle following the impact like in Figure 9 or another type of droplets show steady shapes like in Figure 7, however with flat bottoms. As for the first case, it occurs when the radius is much larger than the capillary length, κ. A force is exerted on the two sides of the surface area of the puddle resulted from the gravitational pressure represented by $\rho g h^2 / 2$. Then Aussillous & Quere et. al. [14] arrived to the conclusion that the thickness of the puddle, h=2κ^{-1}. Once the drop diameter is sufficient that it exceeds the capillary length, then the puddle height is essentially independent. However the volume of the droplet has two formulas; $\pi l^2 h$ and $\frac{4}{3}\pi r^3$. The first is the way of calculating the volume of the droplet in the puddle shape. The second is the way of calculating the droplet in the spherical shape. Equating these two formulas to each other will give the length of the droplet in contact with the powder, l. $l = \sqrt{\frac{2}{3}} r^{3/2} \kappa^{1/2}$.

As for the second case where the droplet stands on top of the powder without forming the puddle as in Figure 9, the bottom surface of the marble is flat, see Figure 11. Quere et. al. [14], estimates the length of this contact area

using the following scaling argument. This case happens when the radius of the

droplet is much less than the capillary length i.e. when R<<κ. Quere et. al. [14],

mentioned that this flattening results from gravitational forces, i.e. the marble

deforms under its own weight. The pressure difference between the drop and

the ambient atmosphere is given by: $\Delta P = \dfrac{2\sigma}{r}$.

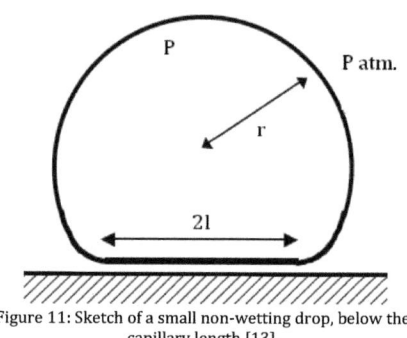

Figure 11: Sketch of a small non-wetting drop, below the capillary length [13].

Since the   deformed area

of the marble is known, $A = \pi l^2$, then a simple force balance yields the following:

$$\Delta P \pi l^2 = \dfrac{4\pi}{3} \rho g r^3$$

, which can then be simply rearranged to give the length of the

contact zone as $l = \sqrt{\dfrac{2}{3} r^2 \kappa}$.

2.3 Formation of Liquid Marbles

Figure 12: Liquid Marble [15]

In this image shown in figure 12, the liquid marble was produced by rolling, which is the crude method that has been presented in previous works [16] to produce full encapsulations. However, it is possible to produce full marbles by simply depositing a liquid drop on a powder surface, if the conditions are favorable for solid-over-liquid spreading to occur. One such example was presented by [16]. Spreading amount over a droplet always depends on the spreading coefficient $_{SL}$ illustrated by equations earlier and show how the works of adhesion and cohesion result in this coefficient. The spreading coefficient however does not predict the behavior of the powder on the surface of the droplet. It can only predict how much of the droplet is coated by the powder. Adding to the equations listed earlier regarding the works of cohesion and adhesion and the solid over liquid spreading coefficient, the following equation represents the spreading coefficient regarding the polar and non-polar intermolecular interactions:

$$S_{/L} = 4\left[\frac{\sigma_S^d \sigma_L^d}{\sigma_S^d + \sigma_L^d} + \frac{\sigma_S^p \sigma_L^p}{\sigma_S^p + \sigma_L^p}\right] - 2\sigma_S$$

The ratio of the solid/liquid interface area and the liquid surface area it replaces is defined according to the following equation:

$$\beta = 1 + (h/r)^2$$

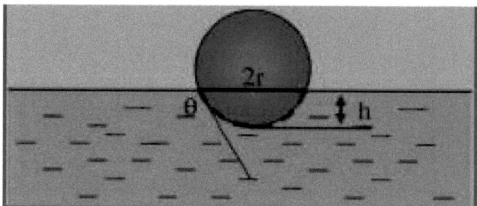

Figure 13: A non-wettable powder particle contacting a liquid surface. The area of the solid/liquid interface (broken line) and the area of the liquid surface replaced by the particles (thick solid line) as it spreads on the liquid surface are different [16].

According to the experiments performed by Nguyen et al. [16], this ratio is almost always between 1 and 2, where β=1 represents the limit of non-wetting and β=2 represents Cassie-Baxter effect. As the particle sizes of the powder decrease, the inter-particle attraction forces between those particles become stronger, and therefore the particles tend to move over the liquid surface.

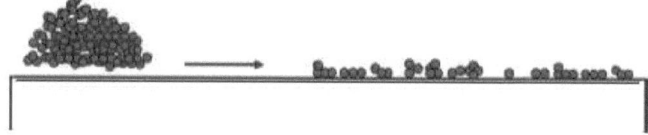

Figure 14: A schematic of a solid powder aggregate disintegrates and expands over a liquid surface [16].

In another research study conducted by McEleney [17], it was reported that the temperature gradient could be one of the major factors affecting the movement of the powder over the liquid surface. The assumption was that the

top part of the drop is exposed to the light for experimental investigation while

imaging, so the temperature at the top of the drop will be higher than that at the

bottom. Thus this temperature difference could drive the motion of the particles

over the surface of the drop [17].

The formation of liquid marbles has been examined extensively by

Hapgood and co-workers [11, 13, 17, 18, 19] and they have identified two main

methods responsible for marble formation: The first is due to the difference

between works of cohesion and adhesion due to the spreading of powder all

around the surface of the liquid drop denoted by the solid over liquid spreading

coefficient $_{SL}$ [13]; The second is due largely to impact kinetic energy of the

droplet.

For a given fluid, the impact kinetic energy, $KE_0 = \frac{1}{2} m u_i^2$, where m is the

mass of the liquid drop, has a profound influence on the marble formation due to

the amount of viscous dissipation. For example, pure glycerol ($\mu \approx 1200$ mPa.s)

induces more viscous dissipation than pure water ($\mu \approx 1$ mPa.s), thus when it

comes to the impact dynamics, more kinetic energy is lost in the case of pure

glycerol through viscous dissipation, thus reducing the tendency for the drop to

deform upon impact and therefore restricting the interaction of the drop and the

powder. Generally, the low-viscosity fluids like that of water's or even less are

observed to have a higher propensity to rebound.

Hapgood and co-workers have developed a rule-of-thumb flow-chart for

determining whether or not an impact will lead to the successful formation of a

liquid marble based on empirical data, which is reproduced here in Figure 11.

These criteria give a starting point for selection of materials for investigating liquid marble formations. In addition, Hapgood et al. have performed experiments to examine the efficacy of the formation of marbles due to the impact alone, without manual manipulation or mechanical agitation and deduced the following equation for the percentage coverage of the liquid drop:

$$Coverage\% = A\left(1 \quad e^{\ bE}\right).$$

Where A,b and E are empirical constants. Some example constants determined by Hapgood et al. [18] are given in table 2 below:

Fluid	Maximum Coverage A (%)	Ease of formation b (% J)
Water	95	0.3
20% Glycerol	96	0.13
40% Glycerol	96	0.12
60% Glycerol	94	0.13
80% Glycerol	93	0.1
100% Glycerol	84	0.045

Table 3 values for "A" and "b" for calculating the coverage percentage [18]

The following diagram shows the steps required for liquid marble formation:

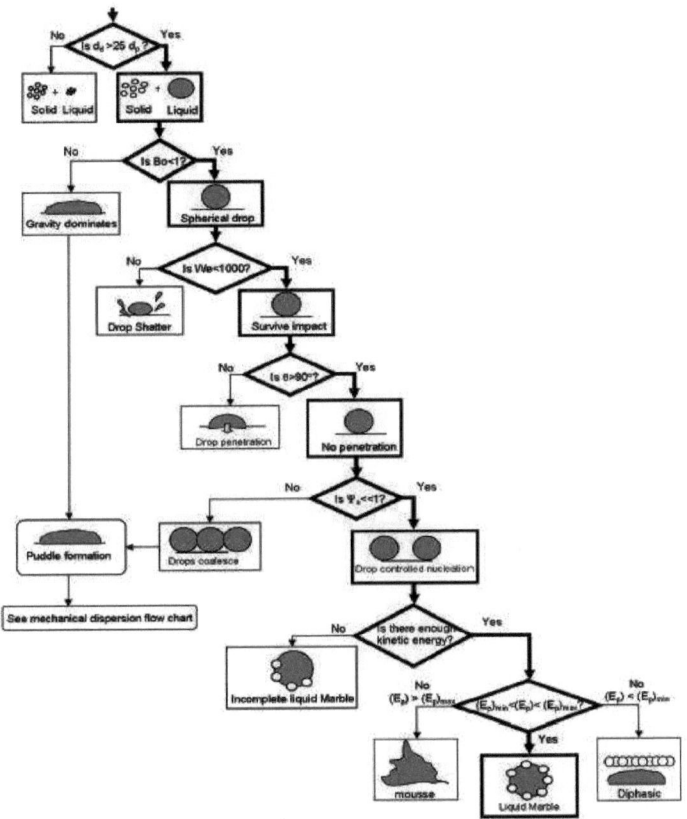

Figure 15: Flow chart steps for liquid marble formation [13]

2.4 Properties of Liquid Marbles

Once a marble is formed, the particles provide the liquid droplet with a protective shell of grains and, as a result, the liquid inside is isolated from further physical contact with other solids or liquids. This then yields some fascinating properties, whereby the marble can deform [14].

Liquid marbles can be fused, divided, caused to spread, burst, or manipulated, through different techniques such as gravitational, electrostatic, or magnetic field [14]. Figure 16 below shows an example of manually dividing a marble into two smaller spherical marbles.

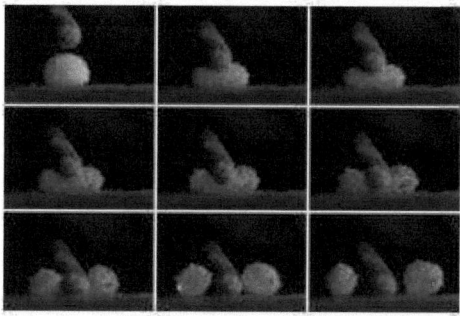

Figure 16: Division of millimetric water marble; images are taken at 4 ms interval. This is an example of manually dividing the liquid marble.

Liquid marbles have many advantages such as having low vapor pressure, ability to withstand wider range of temperatures than that of the liquid alone, i.e. enhanced thermal stability, nonflammability and ionic conductivity [20]. They also have lower melting points than conventional salts. The liquid marbles are, in addition, versatile. They transform from being water miscible to water immiscible and from being hydrocarbon immiscible to hydrocarbon miscible [20]. Also, due to the absence of a contact line, they are in a very mobile state, and a very low frictional force is produced when they are placed on a solid surface, therefore only a very small external force is required to manipulate them. The liquid marbles can thus be used in microfluidic applications and can act as micro-transporters or micro-reactors [14].

Some other characteristics of liquid marbles are that they can float on a liquid surface for some time. The lifetime of a liquid marble with grain sizes between 0.1 and 100 m and a water interior, is approximately 1 minute when placed on a water surface. Normal liquid droplets never survive more than 0.3 seconds [14]. Also, when a marble impacts onto a liquid free surface, it sometimes ejects a small micro droplet as a result of the impact following a mechanism reported by [21]. Liquid marbles can resist shocks and impacts without shattering [14]. Liquid marbles were shown to have the unique property of divisibility without losing their characteristics [14].

Given the unique and fascinating properties of liquid marbles, it is imperative to study their formation from a dynamical perspective – i.e. the initial interaction of a liquid drop onto a hydrophobic powder. This is the basis for the remainder of this thesis.

Chapter Three: Statement of Objectives

The principal goal of this research is to perform a systematic study of the impact dynamics of a liquid drop onto a hydrophobic powder bed, which constitutes the first step in the formation of a liquid marble. To this end, we will employ high-speed imaging with frame rates up to 10000 fps and high optical magnification in attempt to elucidate the fine details of the impact-rebound of milli-metric sized drops. We will also examine the efficacy of marble formation from the simple gravity-driven impact process, without further manual manipulation or mechanical agitation, by using tilted powder beds. We will also perform a comparison of impact of liquid marbles onto solid surfaces with liquid droplets onto hydrophobic solids surfaces. One of the key parameters we will investigate in this study, which has not been the primary focus of previous studies, is the packing fraction. Finally, we will extend our parameter space by examining the impact of micro-droplets, with diameters down to 100 microns, to complement previous studies, which focus mainly on millimetric sized drops.

Chapter Four: Experimental Methods

4.1 Hydrophobizing agent and method

In our experiments, we used a novel method to prepare hydrophobic powders by using a commercially available agent, Glaco Mirror Coat "Zero" (Soft 99 Co., JP). Glaco is an alcohol-based suspension of silica nano-particles. Traditionally, this product is used in aerosol-form, which can be sprayed onto car wing mirrors to repel water droplets. However, the company kindly provided us with a pure solution form, which we use to yield hydrophobic powders.

There are several steps to this process, though; First, the Glaco is diluted with ethanol. This diluted solution is then poured over a sample of the raw powder until the powder is over-saturated at least 2:1 (solution:powder). The wet powder is then placed in a beaker inside an ultrasonic cleaner and sonicated for at least 15 minutes. The powder is then drained and the wet powder mass is then placed in a dessicator, which is evacuated to a low vacuum for at least 2 hours before heat curing in a temperature-control furnace for another 2 hours at 120 C. The powder is thus left with a stable nano-particle coating, which is stable.

4.2 Setup Overview

The setup consists mainly of the high-speed camera and lighting system, a powder bed and a drop delivery system. These tools and materials are set all together as shown in Figure 18. The powder used in these experiments were 0-53 microns, which were treated using the above method. For convenience, the equivalent volume-based mean particle diameter used throughout the results and analysis for the glass beads is $d_b=31$ m. The powder beds were prepared in

a small plexi-glass cylinder with diameter 50 mm and 10 mm depth. In order to
create a flat, reproducible surface, the powder bed was scraped with a straight,
solid edge before testing. In addition, to vary the packing fraction, the bed was
also compressed by pressing down on the surface with a large, flat plexi-glass
plate. Using this preparation method, we were able to produce packing fractions
in the range 0.5<f <0.6.

The following table, Table 4 shows the liquids used in the experiments and
their physical properties. The glycerol solutions indicate percentage by mass
mixed with water, while the SDS solution was made from an aqueous stock
solution of 20% w/w [24].

Liquid	Density (kg/m3)	Surface Tension (mN/m)	Viscosity (mPa.s)
Water	996	73	1
3% SDS	996	38	1
25% Glycerol	1074	69	2.3
40% Glycerol	1110	68.75	4.57
60% Glycerol	1169	68.5	14.2
70% Glycerol	1194	67.8	30.5
80% Glycerol	1219	67.1	78.2

Table 4: Properties of used liquids in the experiments

All experiments were performed at the ambient temperature of $22\ °C$
and relative humidity 55%. The liquids were chosen in order to systematically
vary the viscosity and surface tension of the impacting drop. The droplets are
released from various heights ranging from 1 cm up to 70 cm resulting in
different impact velocities, where $u_i = \sqrt{2gh}$. However, the exact impact velocity
was measured directly from the video sequence. Two sets of experiments are
performed for every liquid. First, we perform normal impacts, where the drop
impacts vertically onto a horizontal powder surface. Second, we perform non-

normal impacts, where the drop impacts vertically onto a titled powder surface. In each case, the rebound is approximately normal to the powder surface.

A 10 l Hamilton syringe is connected through a tube to a needle almost 0.5 mm in diameter. The resulting droplets have a diameter range of 0.6<D<2.3 mm, depending on the liquid physical properties. The needles with a range of exit diameters are produced using glass-puller (Shutter instruments). They are usually blocked at the end. Their tips are broken to produce the droplets. The difference in the diameters of the droplets depends on the breaking of the tip. The drop diameters are also measured directly from the video sequences after performing a spatial calibration (see below).

Figure 17: The glass needles used to pinch off the droplets.

4.3 Video Capture and Analysis

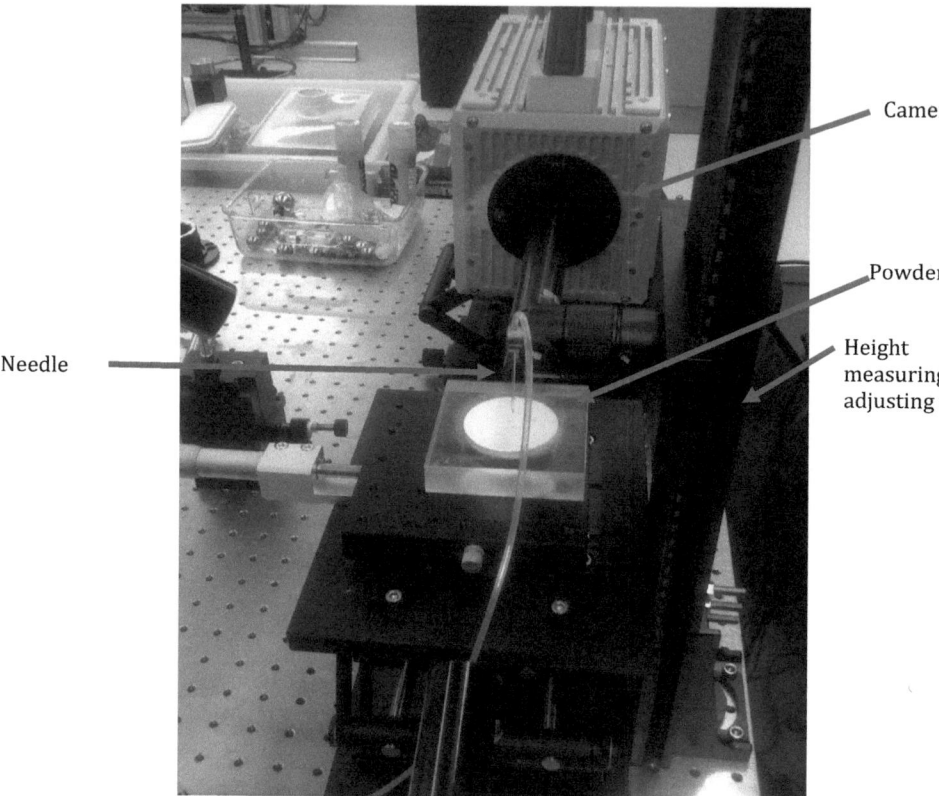

Figure 18: A closer view of some parts of the setup

Figure 18 shows a close-up of the setup where the needle is located just above the powder bed. The height of the nozzle is adjusted by a glider mounted to a vertical optical rail.

4.4 Instrumentation

4.4.1 High-Speed Camera

A Photron Fastcam SA-5 high-speed camera was used that can record up to one million frames per second. However, most of the video sequences for these experiments are recorded at 7000-10,000 frames per second with an effective sensor area of 960x416 Pixels. The total video sequence is cropped to the frames of interest, which means several frames before impact until the drop/marble comes to complete rest. Note that the recording time was not always sufficient to capture the full drainage of the liquid drop into the bed. The recording duration was sufficient to allow for manual triggering. From the video sequences, it was then possible to extract basic data such as the drop size, drop impact speed, maximum spread diameter, maximum rebound height and the times associated with these features. Most of these features were determined manually using Photron Fastcam Viewer (PFV) Software.

Figure 19: SA-5 high-speed camera

4.4.2 Objective Lens

The objective lenses used in the experiments were either a Leica Z16 APO, with a 1.6x mounting lens, or a Mitutoyo 5x microscope objective, fitted to the camera via a c-mount plate. The conversion of the 20 micron sensor pixels into metric measurements, for different magnifications were measured with a calibration plate and are shown in Table 4:

Eye piece magnification zoom	1 pixel in micrometers
0.5x	21.92
0.8x	15.62
1.0x	12
1.25x	10
1.6x	7.8
2.0x	6.25
2.5x	5
3.2x	3.9
4.0x	3.12
5.0x	1.98
6.3x	1.5625
8.0x	1.358

Table 5: Converting the pixels into micrometers according to the magnification of the used lens

4.4.3 Light sources

Due to the short exposure times used to capture the events at high-speed, we require a high-intensity light source. As such, we employed two Sumita 350W metal Halide lights sources equipped with fibre-optic light guides. One light was placed directly opposite the camera, as with conventional silhouette imaging techniques, whilst the other was directed down on top of the powder bed, to help identify the surface of the bed.

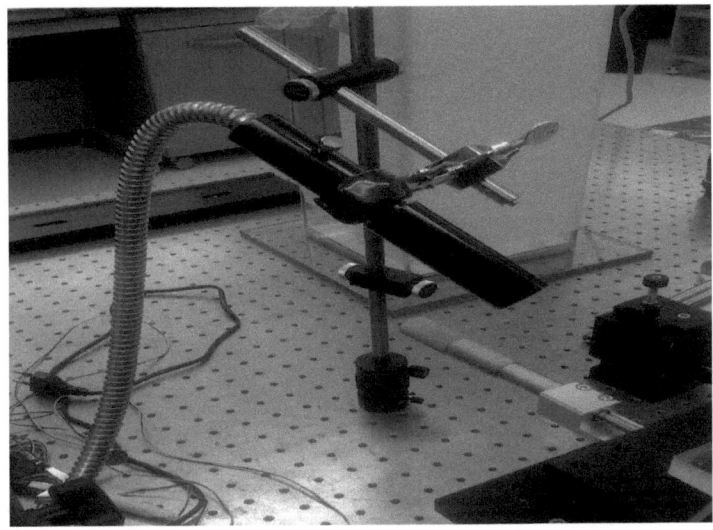

Figure 20: Halide light source directed towards the surface of the impact

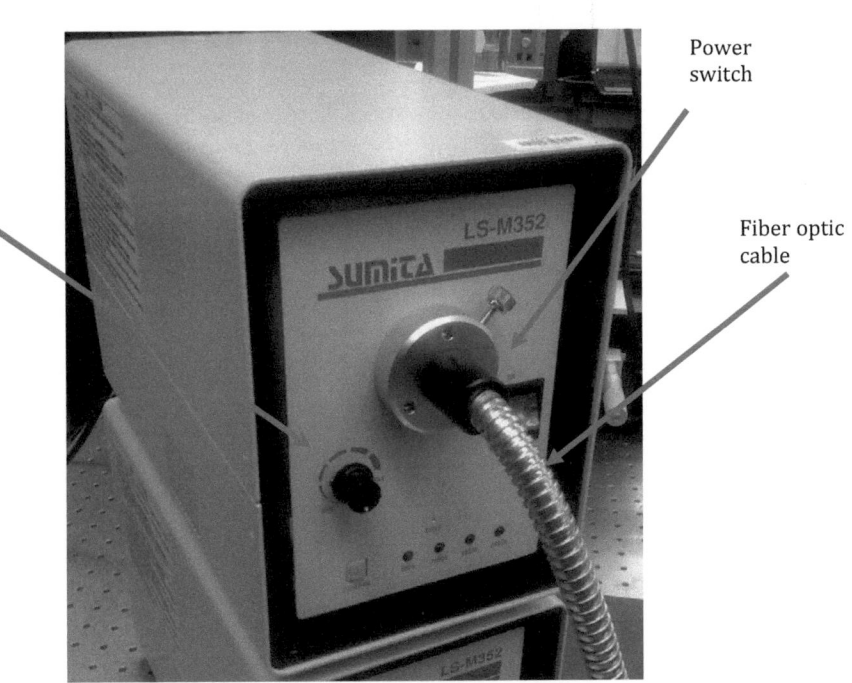

Chapter Five: Results

5.1 Example Impact Sequences

Some typical examples of normal rebounding are shown in Figures 21-26. They are covering some examples of the various liquids, which are 3% SDS, 25% glycerol, 40% glycerol, 60% glycerol, 70% glycerol and 80% glycerol. Some other experiments were done using pure water. The droplets are around 1 mm in diameter.

5.1.1 Horizontal Surface

5.1.1.1 Example 1

Figure 21 shows drop impact at very low impact velocity. Three rebounds can be identified for this particular realization, but the last rebound is very small. The rebounding is likely due to a confluence of the hydrophobicity of the powder and the low surface tension of the drop, which allows for large deformations and oscillations, even for this very low impact velocity. Note that the powder only covers about a 45 degree hemisphere at the bottom of the drop.

1st rebound

nd rebound

3nd rebound

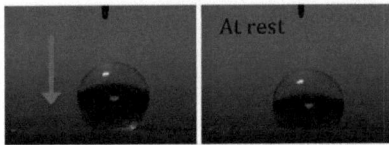

Figure 21: Example sequence for the impact of a 3% SDS droplet, where 3 rebounds were observed onto the powder base before it came to rest, D=1.28 mm,, u_i=0.1 m/s, φ =0.59. We=0.36, Oh=0.0045 and Bo=0.42. Frames shown are taken from the high-speed video sequence at -5, 0, 2.5, 5, 15.7, 20.5, 30.45, 40.4, 41.3, 42.2, 42.5, 42.8, 43.4 and 45.8 ms from the first contact.

5.1.1.2 Example 2

Figure 22 shows a faster impact of a 3% SDS drop at an impact speed of 0.78 m/s (We=29.84, Oh=0.0038 and Bo=0.88). Here the drop rebounds more than four times, in which, the first rebound is 3.81 mm high. Note that the powder now covers almost the entire surface of the drop. Comparing this to Figure 25, where the 25% glycerol drop (We=6.97, Oh=0.0089 and Bo=0.13) impacted at almost the same speed (0.7 m/s) rebounds only once to a height of 2.13 mm. The rebound height is only about half, due to the larger viscous dissipation for the glycerin drop.

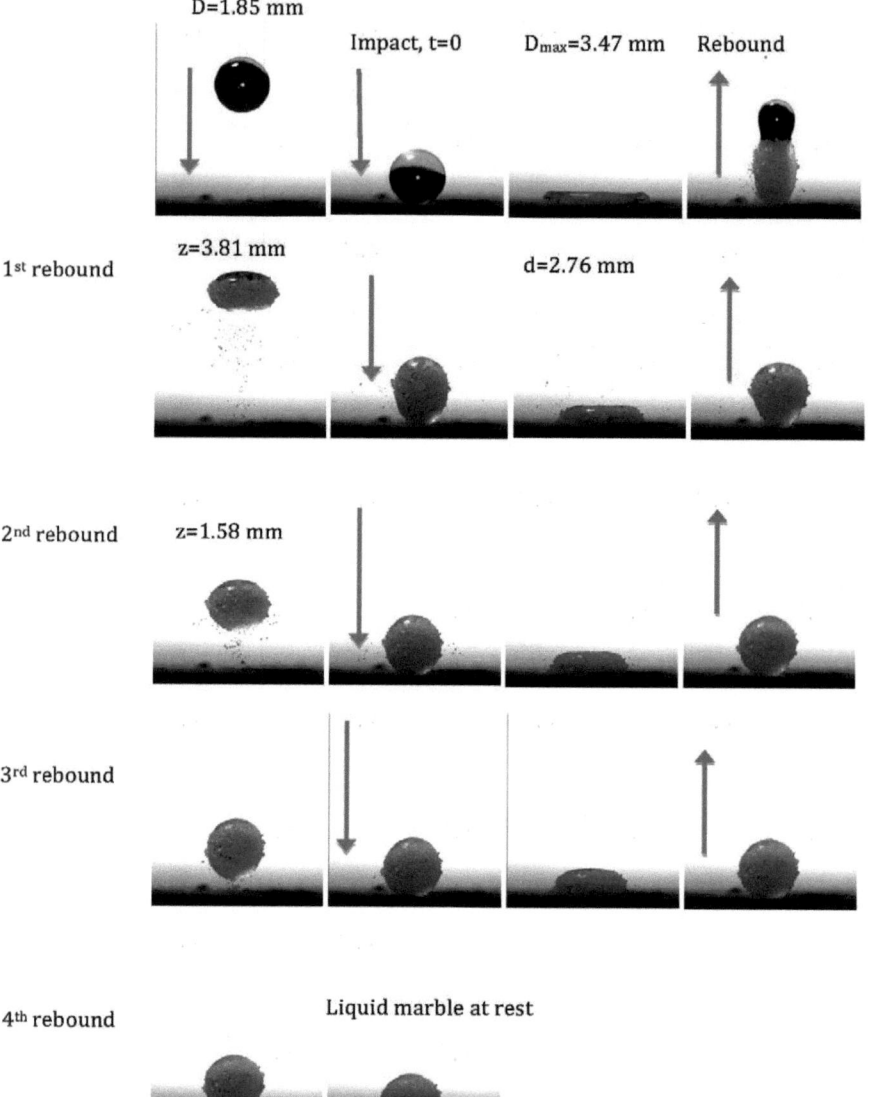

Figure 22: Example sequence for the impact of a 3% SDS droplet showing 4 full rebounds onto the powder base, D=1.85 mm, u_i=0.78 m/s, φ=0.59. We=29.84, Oh=0.0038 and Bo=0.88. Frames shown are taken from the high-speed video sequences at -3.8, 0, 5.55, 11.1, 40.85, 62.5, 68.85, 75.2, 91.8, 108.4, 113.93, 122.23, 130.53, 141.6, 147.13, 155.43, 160.96 and 172.03 ms from impact.

5.1.1.3 Example 3

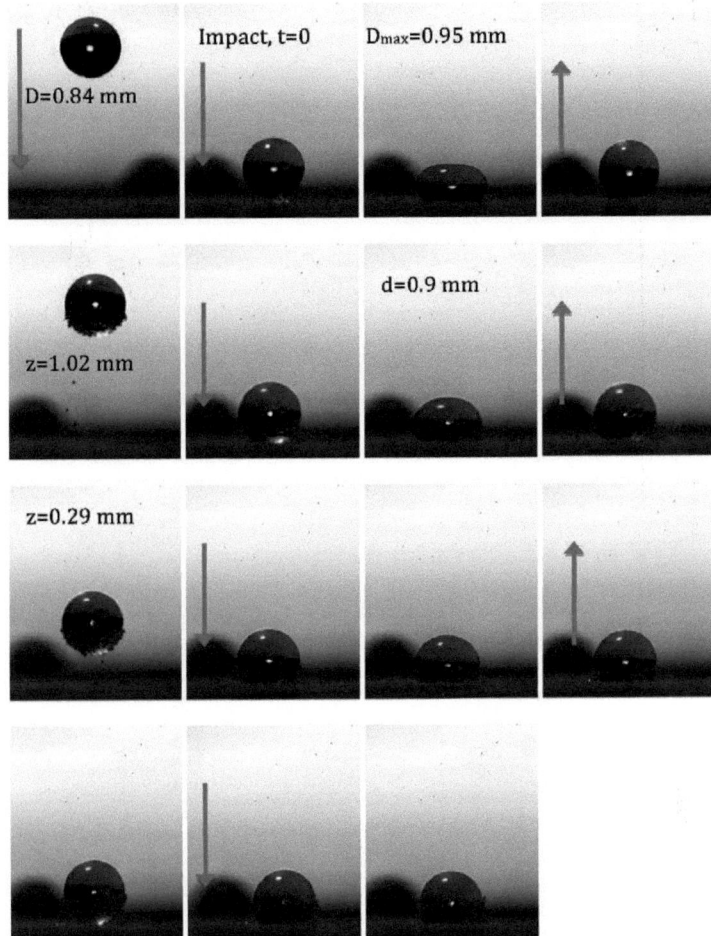

Figure 23: Example sequence for the impact of a 40% glycerol droplet showing three rebounds onto the powder base, D=0.84 mm, u_i=0.2 m/s, φ=0.53. We=0.55, Oh=0.018 and Bo=0.11. Frames shown are taken from the high-speed video sequences at -9, 0, 1.4, 2.8, 17.3, 32, 33.7, 35.4, 43.7, 52, 54.2, 57.4, 59.6, 65.4 and 97.8 ms from impact.

5.1.1.4 Example 4

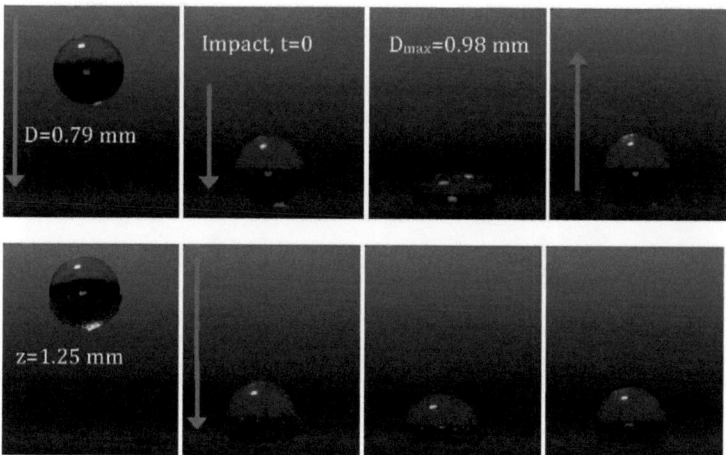

Figure 24: Example sequence for the impact of a 70% glycerol droplet, where only one rebound was observed, onto a base of glass beads, D=0.79 mm, u_l=0.53 m/s, φ=0.59. We=3.95, Oh=0.12 and Bo=0.11. Frames shown are taken from the high-speed video sequences at -2.1, 0, 1.15, 2.3, 19.1, 35.2, 37.2 and 51.9 ms from impact.

5.1.1.5 Example 5

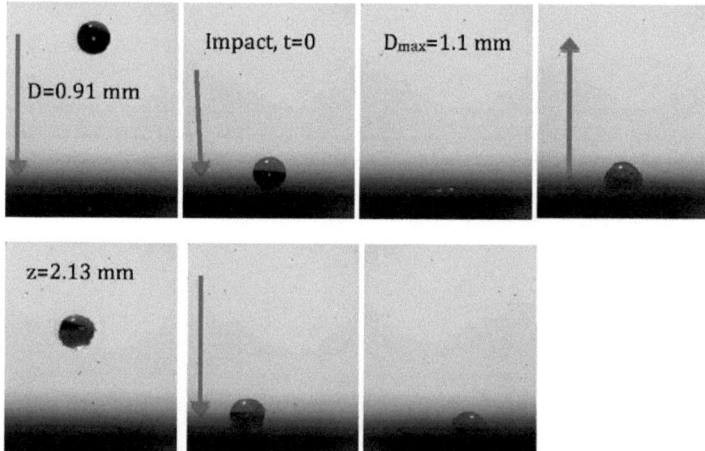

Figure 25: Example sequence for the impact of a 25% glycerol droplet, where only one rebound was observed onto a base of glass beads, D=0.91 mm, u_l=0.7 m/s, φ=0.53. We=6.97, Oh=0.0089 and Bo=0.13. Frames shown are taken from the high-speed video sequences at -5.5, 0, 1.4, 2.8, 25.9, 48.4, 50.6 and 61 ms from impact.

5.1.1.6 Example 6

Figure 26: Example sequence for the impact of an 80% glycerol droplet where only one rebound was observed onto the powder base, D=0.85 mm, u_i=0.44 m/s, φ=0.58. We=3.04, Oh=0.3 and Bo=0.13. Frames shown are taken from the video sequences at -2.7, 0, 1.3, 3.3, 16.4, 29.4 and 47.6 ms from impact.

The normal impacts as shown in Figures 21-26, results in the puddle- or pancake-shape when the droplet reaches maximum deformation (see figure 7). Then the droplet retracts and starts rebounding when it exceeds a certain Bond number, as shown in the below analysis. In some cases the drop oscillates depending on the surface tension, so in cases where the surface tension is low as in the 3% SDS liquid, the droplet deforms more than that for some glycerol solutions. This also depends on viscosity, as viscous dissipation would dampen the oscillation. The drop then rebounds to its highest vertical location. The maximum height of rebound decays with each successive rebound due to energy dissipation. After the first rebound, some sand particles adhere to the base of the drop. If the impact speed is very low, the percentage of the drop covered by the sand is low as well. Whenever the impact speed increases, the percentage coverage increases subsequently. This is due to the maximum spreading of the droplet (D_{max}). As D_{max} increases, more area at the bottom of the drop is in contact with the underlying granular media, picking up more of the powder. Therefore, the coverage of the drop by sand increases. After the rebounds are over, the drop sits on top of the powder and never penetrates into the underlying bed, at least for couple of days. See the results of [11].

More viscous fluids rebounded less in terms of number of rebounds and heights of rebounds. Figure 23 shows the impact of a 40% glycerol drop (We=0.55, Oh=0.018 and Bo=0.11) with a speed of 0.2 m/s that resulted in three rebounds, first of which had a rebound height of 1.02 mm. On the other hand, Figure 26 shows the impact of an 80% glycerol drop (We=3.04, Oh=0.3 and Bo=0.13), with an impact velocity that is more than twice that of the 40% drop (0.44 m/s), where the drop rebounded only once with a height of 0.95 mm, which is slightly less than that of the 40% drop by 0.07 mm.

In the following chapter we will look more systematically at the influence of the various parameters which we have simply shown randomly in the previous figures, where drop size, impact velocity and liquid properties are varied from one figure to the next. But we will also look at oblique impacts, in the next subsection, which we generate by putting the granular beds at an angle.

5.1.2 Inclined Surface

5.1.2.1 Example 1

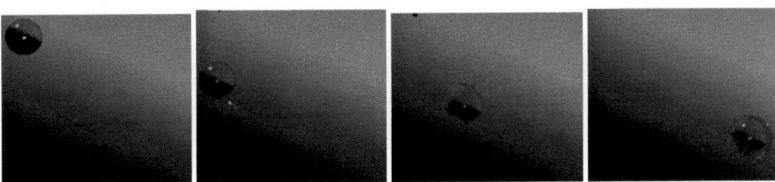

Figure 27: Example sequence for a water droplet impacting onto and rolling over a 24.4° inclined surface of glass beads, D= 0.7 mm, u_l =0.14 m/s, φ=0.55. We=0.19, Oh=0.0045 and Bo=0.067. Frames shown are taken from the video sequences at -9.4, 0, 40.7 and 79.2 ms from impact.

5.1.2.2 Example 2

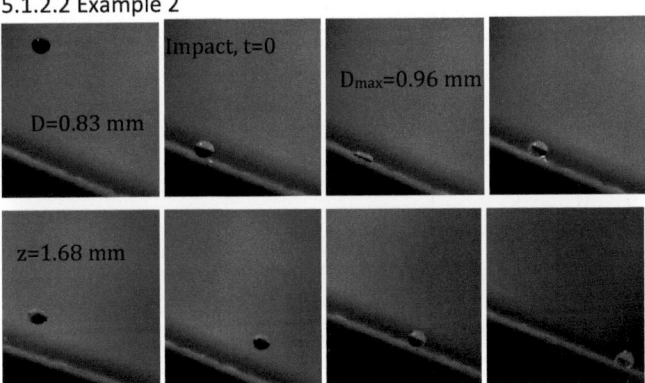

Figure 28: Example sequence for the impact of a water droplet rebounded and rolled over a 24.4° inclined surface of a glass beads, D=0.83 mm, u_l=0.48 m/s, φ=0.58. We=2.62, Oh=0.0041 and Bo=0.093. Frames shown are taken from the video sequences at -13.2, 0, 1, 3.9, 18, 35.4, 43.7 and 74.3 ms from impact.

In Figure 27 the drop impacting at low velocity onto an inclined granular surface. It does not experience rebound, but rolls down the inclined surface, attaining a powder coating, on the back-side, as it rolls. In Figure 28, however, the drop spreads initially and then rebounds, in a similar fashion to normal impacts (e.g. Figures 21-26), before rolling along the surface to form a marble.

5.2 Quantitative Analysis

5.2.1 Spreading Dynamics

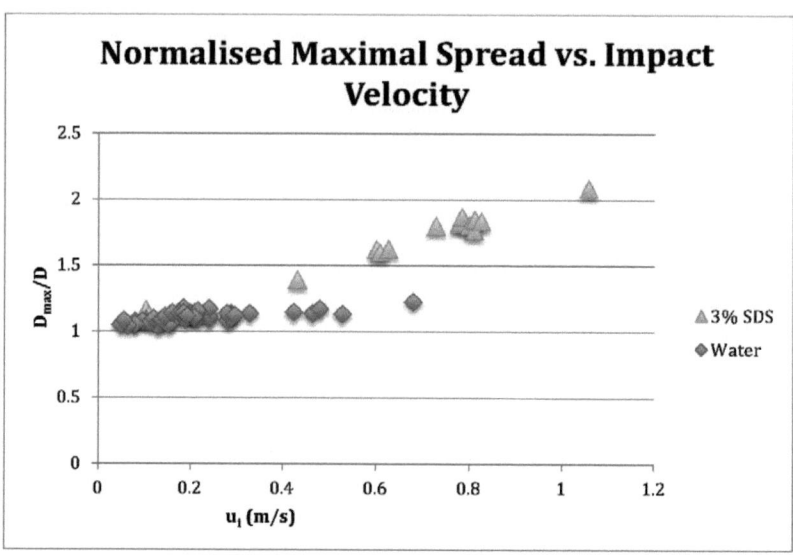

Figure 29: Maximum spread diameter (normalized by initial drop diameter) plotted against impact velocity, u_i, for liquids with low densities and viscosities; $\rho \leq 1000$ and $\mu \leq 1$. The solids packing in the base were 0.54 for water and 0.59 for the 3% SDS.

Figures 29-31 show the relation between the impact velocity or the impact

Weber number and the maximum spread of the drops. The relation is almost

linear in most of the cases, whereby the majority of the data collapses

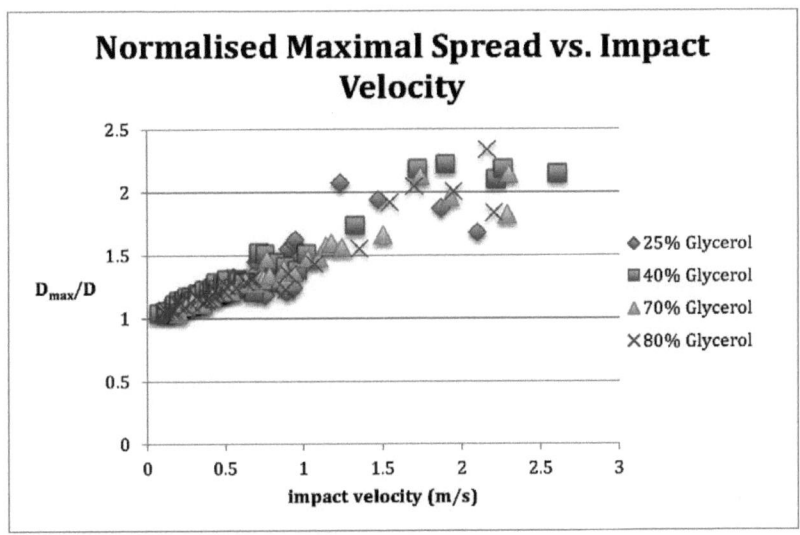

Figure 30: Maximum spread diameter (normalized by initial drop diameter) plotted against impact velocity, u_i, for liquids with $\rho \geq 1000$ and $\mu \geq 1$. The solids packing in the base were 0.53 for 25% glycerol and 40% glycerol, 0.59 for 70% glycerol and 0.58 for the 80% glycerol.

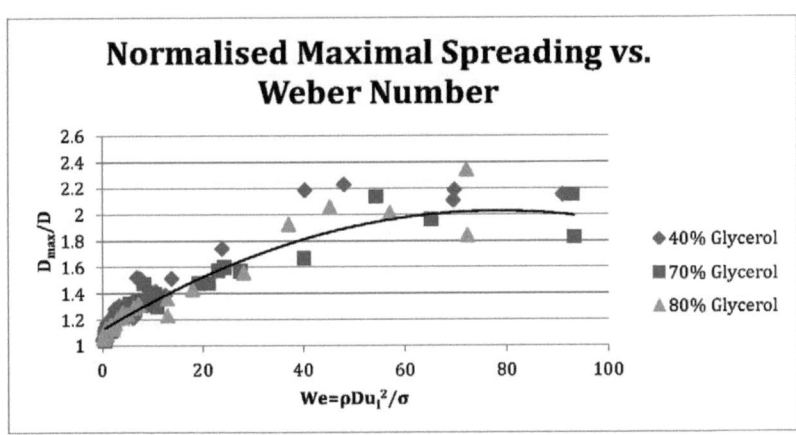

Figure 31: Maximum spread diameter (normalized by initial drop diameter) plotted against impact Weber number, $We = \rho D u_i^2 / \sigma$. The solids packing in the base varied slightly between different fluids. For 40% glycerol, $\varphi = 0.53$, 70% glycerol, $\varphi = 0.59$ and 80% glycerol, $\varphi = 0.58$.

when scaled using $D_{max}/D \sim We^{1/5.5}$. This fit was determined by regression

analysis in Excel. Previously it was shown that $D_{max}/D \sim We^{1/4}$ for $Bo > 1$ and

$D_{max}/D \sim We^{1/2}$ for negligible surface tension [1,24]. The dependency is very clear at the first portion of each chart. After a certain impact velocity or Weber number, the spreading of the drop deviates from its previous trends.

Same is true in Figure 31, where we have for almost all fluids, a dependent relation that stops after the Weber number exceeds 30. In Figure 31 it looks like the spreading of the drops does not really increase, but rather remain constant about the value 1.5 in most of the cases. However the clear independence on impact velocity comes after the value 2.

Figure 29 shows the difference between water and 3% SDS with regards to the spreading deformation, where the surfactant lowers the surface tension allowing more spreading. Figure 30 and 31 shows the difference between the spreading for low-viscosity fluids such as 3% SDS (μ=1 mPa.s) and 25% glycerol (μ=2.3 mPa.s) and the values for higher viscosity fluids such as 60% glycerol (μ=14.2 mPa.s). The 60% glycerol shows less spreading at the same impact velocities of the other two fluids. These results agree with [11], in which Marston et al. mentioned that the lower surface tension liquids deform more than the higher ones and the variation in D_{max} regarding the viscosity is due to the viscous dissipation of energy during impact.

5.2.2 Impact time

The following figures will show the normalized contact time of the drop during the impact and rebound from the surface.

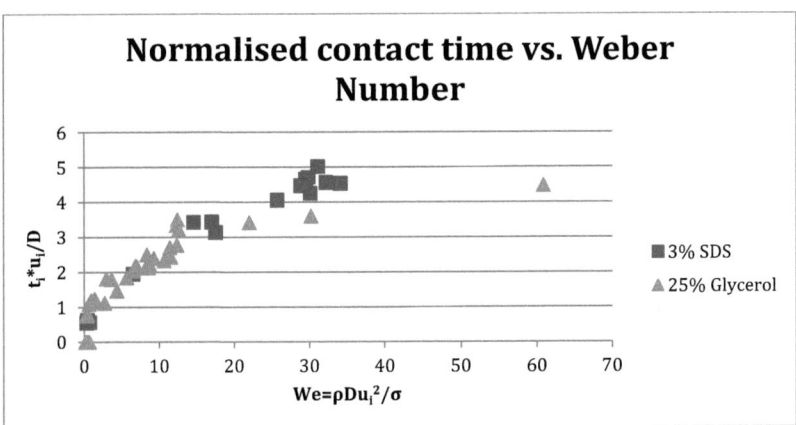

Figure 32: Contact time from impact to the start of the first rebound plotted against the Weber number, We=$\rho Du_i^2/\sigma$. The solids packing in the base varied slightly between both fluids. For 3% SDS, $\varphi=0.59$ and for 25% glycerol, $\varphi=0.53$.

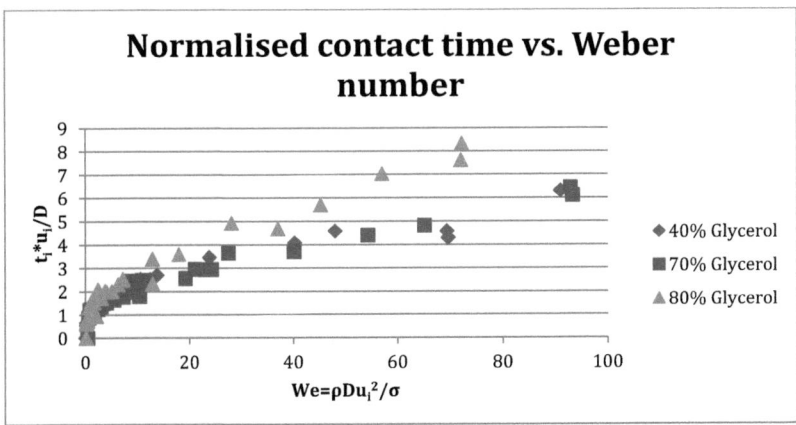

Figure 33: Contact time from impact to the start of the first rebound plotted against the Weber number, We=$\rho Du_i^2/\sigma$, for higher liquid viscosities. The solids packing in the base varied slightly between the three fluids. For 40% glycerol, $\varphi=0.53$, for 70% glycerol, $\varphi=0.59$ and for 80% glycerol, $\varphi=0.58$.

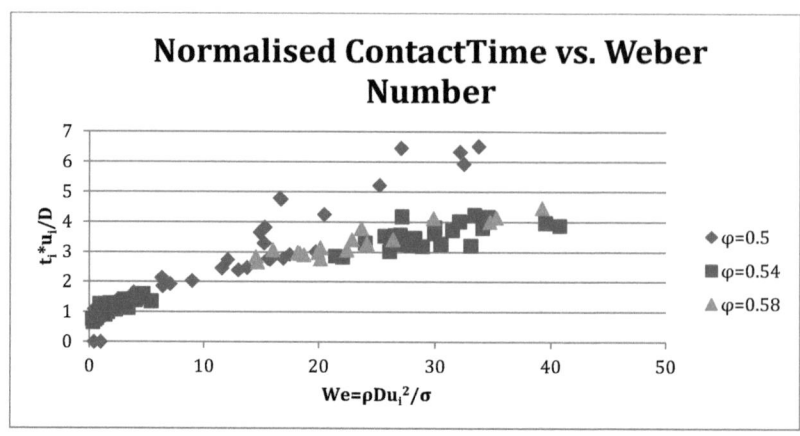

Figure 34: Normalized contact time from impact to the emergence of the drop, plotted against the Weber number, $We=\rho D u_i^2/\sigma$, for different packing fractions of the granular beds, in the range $0.5<\varphi<0.58$.

According to the former studies conducted regarding the impacting onto powder, the following relation is expected, $t_i u_i/D \sim We^{1/2}$. The previous charts, Figures 32 and 34, agree with the following relation, $t_i u_i/D \sim We^{1/2.5}$, which is very close.

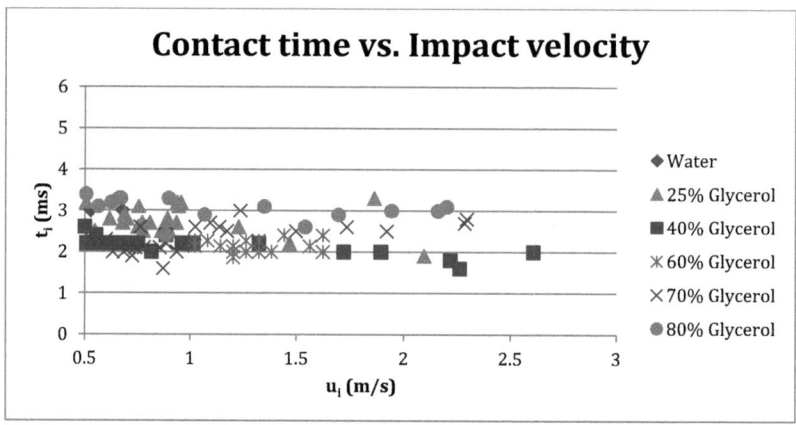

Figure 35: Contact time is plotted against the impact velocity for various liquids.

According to [25], the impact velocity does not have any effects on the impact (contact) time, which agrees totally with the previous chart, Figure 35.

5.2.1.3 Rebounding

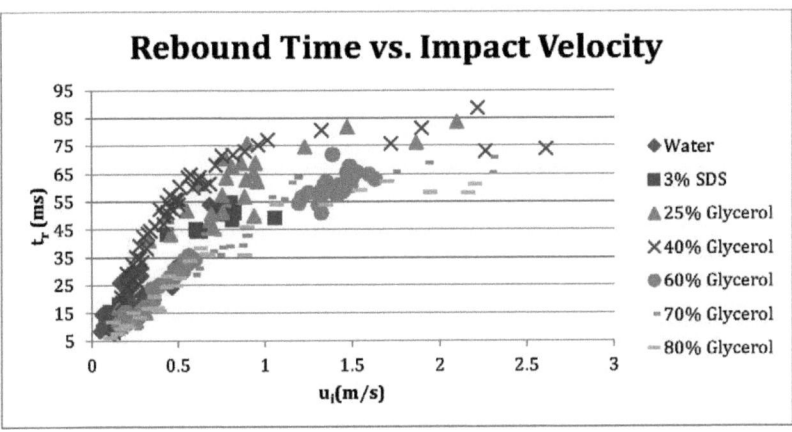

Figure 36: Rebound time from impact to rebound is plotted against the impact velocity for different liquid viscosities. The corresponding packing fractions are 0.54, 0.56, 0.53, 0.53, 0.54, 0.59 and 0.58 according to the sequence of the liquids in the chart respectively.

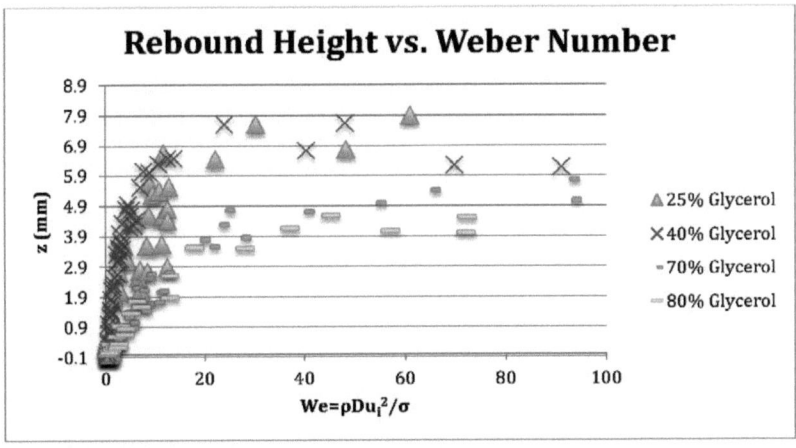

Figure 37: Rebound height is plotted against the impact Weber number at the following packing fractions of each liquid: 0.53, 0.53, 0.59 and 0.58 according to the sequence of the liquids listed in the chart respectively.

Figures 36 and 37 show the effects of impact velocity and We numbers on the rebound time. For all the cases, the rebound time and height are dependent on the impact velocity/Weber number up to a certain upper limit, and then they behave independently. In Figure 36, the limit for the impact speed is 1 m/s. Then the rebound time is almost constant with $t_r=75$ ms. Figure 41 shows a comparison between four different types of liquids. The more viscous the fluid is, the less rebound height it reaches. This is due to the viscous dissipation losing kinetic energy during the impact. Figure 37 explicates the relation between the Weber number and the rebound height. At Weber number (We=20), the data start behaving differently. So the relation can be considered linear till We number around 10, when a transition occurs moving from linearity to the maximum value for the rebound height which never exceeded 8 mm for 25% and 40% glycerol. The 70% and 80% exhibit lower rebound heights for their maximum limits. The maximum for the 70% was 6 mm and the maximum for the 80% was 5 mm. This indicates that for the same conditions of the packing, impacting and surrounding environment, the lower viscous flows exhibit more rebounding heights than the higher ones, as less of the kinetic energy is lost to viscous dissipation and is instead stored in the surface energy and then released during the rebounding of the drop and enters back into upwards kinetic energy, by the surface tension generating pressure gradient inside the drop, which pushes it back up from the granular bed.

5.2.2 Inclined Surface

The results show that impact onto sloping surfaces are close to those for the horizontal plane. The rebound time increases directly proportional with the impact velocity. The relation is almost linear. The different fluids showed close

results where the properties did not contribute to much difference as shown in

Figure 38. More packed bed resulted in more rebounded height eventually.

Generally for the different concentrations, the results show that there is a critical

Weber number (We=1) at which the droplets start rebounding as shown in

figures 39 and 40. Figure 39 is concerned with the rebound height in the vertical

direction, while figure 40 is concerned with the rebound length, which is the

rebound in the horizontal direction (x-direction).

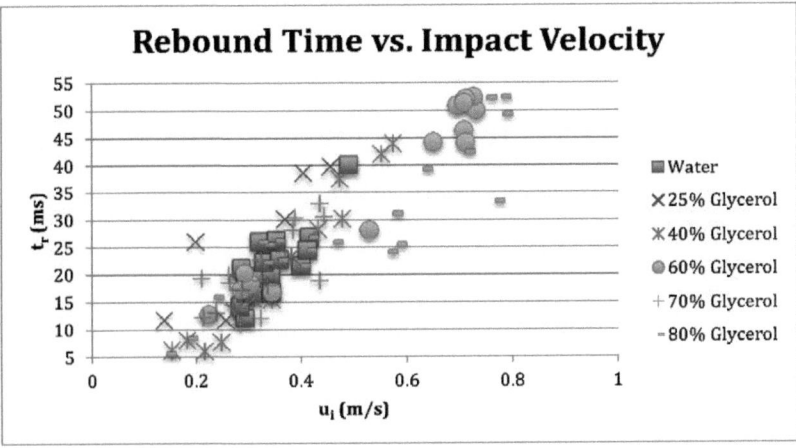

Figure 38: Rebound time is plotted against impact velocity for different liquids, for inclined surfaces. The solids packing of the base varied slightly between repeat trials in the range $0.5<\varphi<0.6$. Target granular surfaces were inclined with angles in the range $22°\leq\alpha\leq30°$.

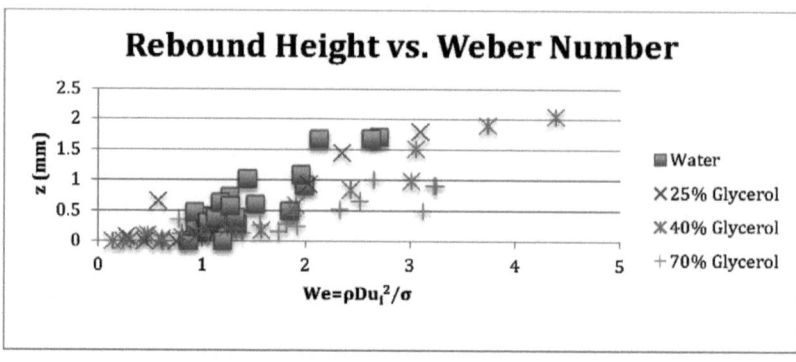

Figure 39: Rebound height is plotted against impact Weber number for various liquids. The solids packing in the base varied slightly between repeat trials in the range $0.5 < \varphi < 0.6$. The target granular surfaces were inclined with angles in the range $22° \leq \alpha \leq 30°$.

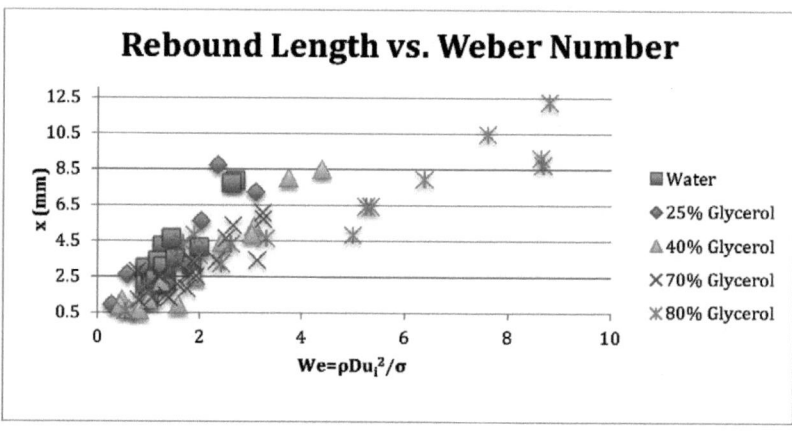

Figure 40: Rebound length is plotted against impact Weber number for variant fluids. The solids packing in the base varied slightly between repeat trials in the range $0.5 < \varphi < 0.6$. Impacting surfaces were inclined with angles in the range $22° \leq \alpha \leq 30°$.

5.3   Splashing and Satellite Drop Generation

5.3.1   Satellite Droplets

Upon impacting at high speeds for some fluids having Bo≥1, some very small droplets get extracted from the original drop and form their own small marbles as shown in the following Figures.

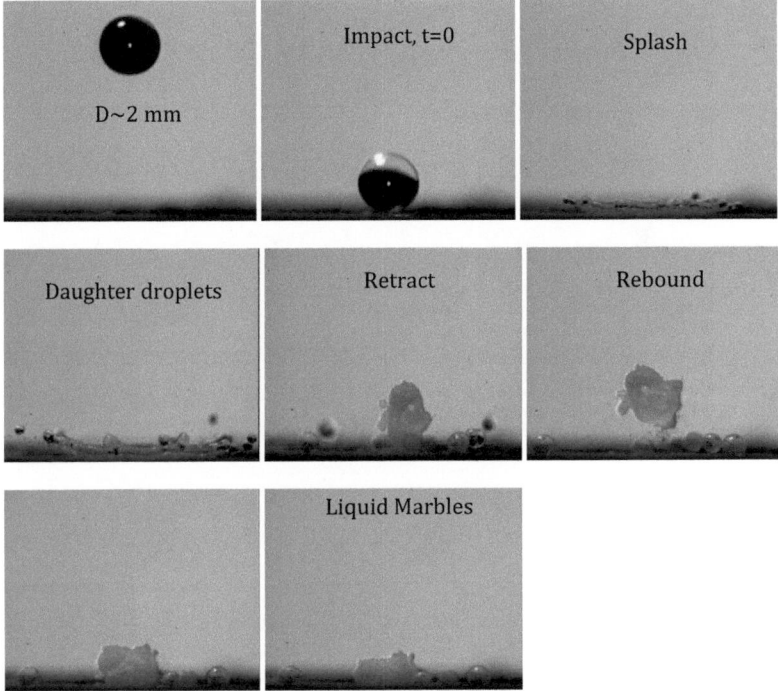

Figure 41: Example sequences for the impact of a 3% SDS droplet rebounded once, D=1.95 mm, onto a base of glass beads, ui=2.082 m/s, φ=0.59. We=221.49, Oh=0.0037 and Bo~1. Frames shown are taken from the high-speed video sequences at -2.1, 0, 1.7, 4.4, 12.2, 25.4, 42.2, 48.2 and 72.1 ms from impact.

Satellite droplets are being emitted at the fourth frame in Figure 41, where this happened as a result of the high Weber number, can also be an important factor. The final image shows liquid marbles formed of different sizes and shapes. The main droplet has formed a deformed shape i.e. non-spherical shape, whereas the satellite drops have formed spherical marbles.

Having inclined surfaces of impacts resulted in the following sequence, shown in Figure 42.

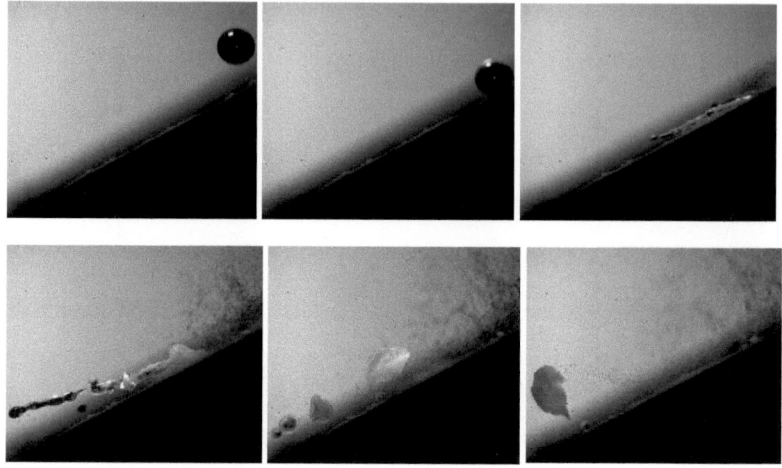

Figure 42: Example sequences for the impact of a 3% SDS droplet on a 27 ° inclined surface, D=1.95 mm, onto a base of glass beads, ui=2 m/s, φ=0.59. We~200, Oh=0.0037 and Bo~1.

5.3.2 Jetting

Some experiments witnessed the emergence of satellite droplets that quench from the top of the original drop. In some cases one satellite droplet emerged. In some others more than one were observed. These resulting droplets are very deformed.

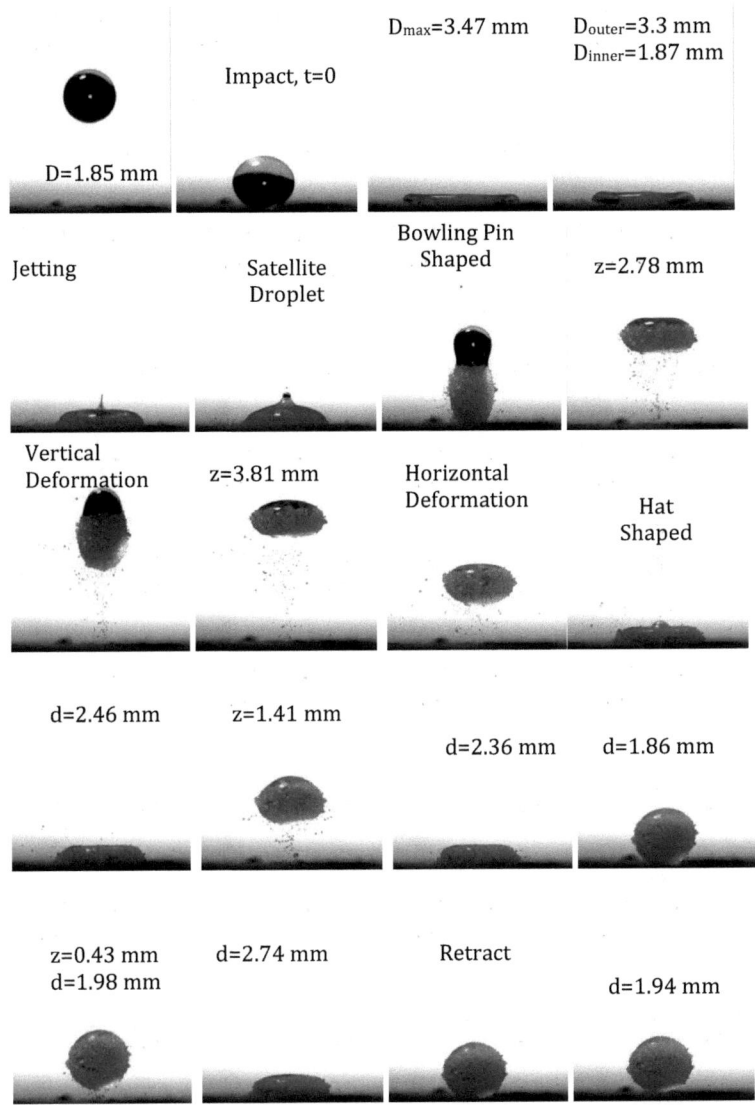

Figure 43: Example sequences for the impact of a 3% SDS droplet rebounded quadruple, D=1.85 mm, onto a base of glass beads, ui=0.78 m/s, φ=0.59. We=29.84, Oh=0.0038 and Bo=0.88. Frames shown are taken from the high-speed video sequences at -3.9, 0, 2.3, 4.4, 5.6, 5.8, 11.4, 33.2, 38.4, 46.4, 52.1, 63, 66.8, 68.2, 75.9, 114.2, 131.7, 151.3, 161.4 and 166.2 ms from impact.

Figure 43 shows the oscillations performed by the 3% SDS drop after its

impact at a speed of 0.78 m/s. The oscillations are believed to happen due to the low surface tension. At 5.8 ms from impact, a satellite droplet is ejected with a diameter of 0.07 mm. Just before the release of the satellite droplet by 1 ms, it is shown at 4.4 ms, that the spreading of the main drop consists of two diameters. One is the outer diameter and the other is inside. Between the time frames 2.3 ms and 5.8 ms, the inner diameter decreases until it diminishes and as it disappears, the satellite droplet is ejected. In the recent study conducted by Marston et. al [11], they were able to produce a pinned shape and it was deduced then that the interaction of the liquid with the top layers of the powder allows a void circular shape to take place resulting in the pinned structure. In our case, the void shape can be the reason behind the creation of the inner diameter that diminishes to zero as the drop gains its energy to rebound.

Chapter Six: Conclusions and Future Work

This thesis work illustrates the novel dynamics of the impacting of drops onto powders. Seven different types of liquids were tested and full sets of experiments were performed on each liquid. The input parameters varied were the diameters of the drops, the impacting speed, the liquid viscosity and surface tension, as well as the packing fraction of the powder. The output parameters studied were the contact time, the rebound time, number of rebounds, rebounding heights, and maximum spread. The range of physical properties led to a range of dimensionless numbers used to quantify the phenomena. Dimensionless numbers taken into considerations were the Bond number, Weber number and Ohnesorge number. Two complete sets of experiments were done on every liquid. One for horizontal powder surfaces, the other for inclined powder surfaces.

The contact time during which the drop deforms and rebounds was shown to be independent of on the impact velocity and a large range of viscosities. The rebound height decreases with increase of the viscosity of the liquid, owing to the viscous dissipation of the impact kinetic energy. Lower surface tension resulted in more deformations, larger spreading diameter and larger shape oscillations, as observed for the water solution of SDS.

In addition to the impacts onto horizontal surfaces, we also investigated impacts onto inclined surfaces, i.e. oblique impacts. Measuring the rebounding heights and lengths, we found that the less viscous fluids rebound more in terms of both. Therefore, the viscous dissipation also plays an important role here.

Future work can be focused on experiments regarding the dynamics of the liquid marbles after their formation. Examples include experiments comparing liquid-marbles impacts onto solid surfaces with drops impacting on hydrophobic surfaces. Some other experiments can study the properties of liquid marbles and their impact on liquids and other different types of powders. In addition, the phenomena, which were discovered through our experiments such as the daughter droplets can be further investigated. Generally, liquid marbles are an ongoing research topic of study which can be broadened and explored more for widening their potential applications numerous areas, such as in the pharmaceuticals industry.

69

WORKS CITED

[1] S.M. Ivenson, J.D. Lister, K. Hapgood, B.J. Ennis, Nucleation, growth and breakage phenomena in agitated wet granulation process: a review, *Powder Technology*, **117**, 3-39, (2001).

[2] A. B. Subramaniam, Abkarian, M., Mahadevan, L .& Stone,H., Non-spherical bubbles, *Nature* **438**, 930 (2005).

[3] J. O. Marston, Drop Impact onto Powder Surfaces, ME seminar series presentations, 2. November 2011.

[4] K. P. Hapgood, B. Khanmohammadi, Granulation of hydrophobic powders, Powder Technology, **189**, 253-262 (2009).

[5] List of Pharmaceutical Companies, Wikipedia, The Free Encyclopedia, March 2010.

[6] H. Katsuragi, Morphology Scaling of Drop Impact onto a Granular Layer, *Physical Review Letters*, **104**, 218-001 (2010).

[7] H. Katsuragi, Length and time scales of a liquid drop impact and penetration into a granular layer, *J. Fluid Mech.*, **675**, 552-573 (2011).

[8] S. T. Thoroddsen, A. Q. Shen, Granular Jets, *Phys. Fluids* 13(1), 4-6 (2001).

[9] C. Clanet, C. Beguin, D. Richard, D. Quere, Maximal deformation of an impacting drop, *J. Fluid Mech.*, vol. **517**, pp. 199-208 (2004).

[10] J. Marston, E. Li, S. T. Thoroddsen, Evolution of fluid-like granular ejectas generated by sphere impact, *J. Fluid Mech.*, **704**, p. 5-36 (2012),

[11] J.O. Marston, S.T. Thoroddsen, W.K. Ng, R.B.H. Tan, Experimental study of liquid drop impact onto a powder surface, *Powder Technology,* **203** (2) 223-236 (2010).

[12] F. Buzsaky, Wet granulation system compressing at least one ultrasonic nozzle, *Patent application publication*, US2011/0287168 A1.

[13] N. Estiaghi, K.P. Hapgood, A quantitative framework for the formation of liquid marbles and hollow granules from hydrophobic powders, *Powder Technol.* (2011), doi: 10.1016/j.powtec.2011.05.007

[14] P. Aussillous, D. Quere, Properies of liquid marbles, *Proc. R. Soc. A*, **462**, 973-999 (2006).

[15] P. Aussillous, D. Quere, Liquid marbles, *letters to Nature*, VOL **411**, 21 June (2001).

[16] T.H. Nguyen, N. Eshtiagi, K.P. Hapgood, W. Shen, An analysis of the thermodynamic conditions for solid powder particles spreading over liquid surface, *Powder Technology* **201** (2010) 306-310.

[17] P. McEleney, et al., Liquid marble formation using hydrophobic powders, *Chemical Engineering Journal* **147** (2-3) (2009) 373-382.

[18] N. Estigashi, J.S. Liu, W. Shen, K.P. Hapgood, Liquid marble formation: Spreading coefficients or kinetic energy?, *Powder Technology*, **196** (2009) 126-132.

[19] T.D. Blake, J. De Coninck, The influence of pore wettability on the dynamics of imbibition and drainage, *Colloids and Surfaces A: Physicochem. Eng. Aspects* **250,** 395-402 (2004).

[20] L. Gao, T.J. McCarthy, Ionic Liquid Marbles, *Langmuir*, **23**, 10445-10447 (2007).

[21] S.T. Thoroddsen, K. Takehara, The coalescence cascade of a drop, *Phys. Fluids* **12**, 1265-1267 (2000).

[22] Tosyl, Wikipedia, The Free Encyclopedia, October 2011.

[23] Heptane, Wikipedia, The Free Encyclopedia, November 2011.

[24] "SDS Solution, Molecular Biology Grade (10% w/v)." *Promega.* Promega Corporation, 2012. Web. 9 Jan 2012.
<http://www.promega.com/products/biochemicals-and-labware/biochemical-buffers-and-reagents/sds-solution_-molecular-biology-grade-_10_-w_v_/>.

[25] D. Richard, C. Clanet, D. Quere, Contact time of a bouncing drop, *Nature* **417** 811 (2002).

[26] Y. Renardy, S. Popinet, L. Duchemin, M. Renardy, S. Zaleski, C. Josserand, M.A. Drumright-Clarke, D. Richard, C. Clanet, D. Quere, Pyramidal and toroidal water drops after impact on a solid surface, *J. Fluid Mech.,* **484**, 69-83 (2003).

Printed by Books on Demand GmbH, Norderstedt / Germany